NO-GRID

Survival

16 DIY Projects for Water, Heat, Shelter & Emergency Power.

Riven Calden

No Grid Survival Project

ISBN: 978 - 1 - 956369 - 30 - 4

DEDICATION PAGE

For every curious mind who believes self-reliance is still a superpower.

And for every family who understands that preparation isn't fear, it's wisdom.

ABOUT THIS BOOK

This book is a practical, hands-on survival manual built for one purpose: **to help you operate confidently when the grid fails**, whether for 24 hours or several months. It's designed for beginners, hobbyists, DIY lovers, and anyone who wants to become more self-reliant.

Inside these pages, you'll learn how to build off-grid systems, improvise tools, process food, generate fire and water, and stay safe in crisis situations. Every project is written in plain language, with step-by-step instructions, diagrams, material lists, troubleshooting notes, and beginner-friendly explanations.

This isn't theory.
This is **real-world, field-tested survival knowledge.**

Whether you're preparing for power outages, economic collapse, natural disasters, or simply want to be more capable outdoors, this book gives you a full toolkit to stay alive, stay stable, and stay ahead.

The goal is simple:
Make you the person who doesn't panic when the lights go out.

HOW TO USE THIS BOOK

This book is structured like a working manual, not a casual read. Here's how to get the most from it:

1. Start with the Basics

The first sections explain survival mindset, essential tools, and foundational skills. Don't skip these, every project later depends on them.

2. Choose Your Projects Based on Your Need

Each project is labeled by category:

- Water Systems
- Heat & Fire
- Shelter
- Food Preservation
- Communication
- Security
- Power & Energy
- Tools & Repair

This helps you jump directly to what you need.

3. Follow the Step-by-Step Format

Every project includes:

- A short explanation (why it matters)
- A clear materials list
- A numbered, easy-to-follow build guide
- Safety notes
- Troubleshooting tips
- Where and when to use it
- Variations and upgrades

If you can follow a recipe, you can build anything in this book.

4. Make Use of the Checklists and Quick Guides

The final section gives you fast-access survival sheets you can use in real emergencies—no page flipping, no guessing.

5. Practice Before You Need It

Survival knowledge is useless if you wait until disaster strikes.

Test a project every weekend.

Build a small kit.

Learn one new method a month.

Consistency builds capability.

6. Personalize Your Builds

You'll find "custom upgrades" at the end of selected projects, showing how to improve your setup based on budget, environment, or available materials.

7. Treat This Book Like a Tool

Dog-ear pages.

Highlight key steps.

Store it somewhere accessible.

In a real grid-down situation, this book may become one of your most valuable possessions.

TABLE OF CONTENT

SECTION 1

INTRODUCTION TO OFF-GRID SURVIVAL

What "No-Grid Survival" Really Means

Going off-grid isn't about living like a caveman or abandoning modern life. It simply **means you can continue functioning when the systems you depend on suddenly stop working**, electricity, water supply, communication networks, fuel stations, ATMs, supermarkets, or transportation.

A grid failure can come from:
- Natural disasters
- Infrastructure overload
- Cyberattacks
- Economic collapse
- Political instability
- Severe storms
- Power grid malfunction
- Widespread supply-chain breakdown

When any of these hit, most people freeze. They panic because they're wired to rely on convenience.

Survivalists don't.

"No-grid survival" means:
- You can **cook without electricity**
- You can **purify water from almost any source**
- You can **build shelter with basic materials**
- You can **protect yourself and your loved ones**
- You can **keep food edible for weeks or months**
- You can **communicate when networks go down**
- You can g**enerate heat, light, and fire**
- You can **repair or improvise tools**

Being off-grid is **simply the ability to continue life as normally as possible when everyone else is stuck.**

The Survival Mindset: Your Real First Tool

Every survival manual begins with gear, but gear won't save you if your **mind collapses under pressure.**

A strong survival mindset includes:

1. Calm Under Stress
Panic makes people waste energy, water, and opportunities.
The calm person makes better decisions.

2. Adaptability
Nothing ever goes exactly as planned during a crisis.
Your ability to adjust determines your chances.

3. Willingness to Work With What You Have
Survival isn't about perfection—it's about improvisation.

4. Situational Awareness
You need to be able to notice things before

they become threats:
changes in weather, strange sounds, structural weakness, unsafe water, smoke, movement.

5. Problem-Solving Skills

You will constantly encounter challenges that require simple engineering, logic, and creativity.

6. Resource Preservation

The amateur uses everything too quickly. The survivor makes every resource stretch.

Before touching any tool or project in this book, you need to commit to developing this mindset.
It's the foundation for everything else.

What Happens When the Grid Fails (and Why You're Preparing)

Most people underestimate how fast society collapses during a prolonged outage.
Within **24 hours** of a major failure:

- ATMs stop working
- Fuel stations shut down
- Food spoils in refrigerators
- Water pumps stop functioning
- Phone networks become inconsistent
- Hospitals start rationing energy

Within **3 days:**

- Supermarkets run out of food
- Clean water becomes scarce
- Waste builds up
- Communication becomes unreliable
- Crime significantly increases in urban areas

Within **7 days:**

- Supply trucks stop moving
- Pharmacies empty out
- People begin relocating, scavenging, or looting
- Medical emergencies become deadly
- Security becomes your most important priority

If you're not prepared, even a small grid failure can ruin your life.
If you are prepared, even a long disaster becomes manageable.

That's why this book exists.

The Four Pillars of Off-Grid Survival

If you can manage these four areas, you can survive practically anything.

Pillar 1: Water
Clean water is life.
You need to know how to find it, filter it, purify it, and store it.

Pillar 2: Fire & Heat
Fire gives you warmth, sterilizes water, cooks food, wards off animals, and signals for help.

Pillar 3: Shelter
A good shelter protects you from cold, heat,

rain, sun, wind, insects, and predators.

Pillar 4: Food & Preservation

Food gives you strength, energy, and morale.

Preservation allows you to store resources long-term.

These four pillars are the backbone of every project in this book.

Everything you learn later will reinforce them.

Essential Skills Every Beginner Should Master First

Before attempting the more advanced projects, make sure you can confidently do the basics:

1. Start a fire using more than one method

Matches are easy.
Lighters fail.
Learn ferro rods, fire plough, char cloth, and spark ignition.

2. Build a basic shelter

Know how to use paracord, tarps, natural materials, and simple knots.

3. Purify water from almost any source

Boiling, filtration, solar distillation, improvised sand filters.

4. Use a knife safely and effectively

Cutting, carving, splitting, scraping, sharpening.

5. Navigate with or without GPS

Natural navigation, landmarks, sun, shadow, and improvised compasses.

6. Cook without electricity

Rocket stoves, solar cookers, open fire methods, clay ovens.

7. Maintain situational security

Basic awareness, safe camp positioning, avoiding danger zones.

8. Improvise tools from scrap or nature

Hammers, spears, hooks, stakes, frames, containers.

These skills are simple, but not optional.

Setting Realistic Expectations

Off-grid living is rewarding, but you must understand the challenges:

- You will get tired.
- Projects take more time than expected.
- Weather can ruin your plans.
- Materials may not always be available.
- Mistakes will happen, even dangerous ones.
- You will waste energy if you push too hard.

But here's the truth:

Every survival expert you admire started as a beginner.

Consistency is what makes you capable.

Your First Step Starts Here

You don't have to be a hardcore prepper.
You don't need expensive gear.
You don't need a large backyard or a bush camp.

Start small:
Build one project.
Practice one skill.
Test one fire-starting method.

This book is designed so that with every chapter, you become more independent, more skilled, and more confident.

By the time you reach the final page, you'll be able to live without the grid—not because you escaped society, but because you learned how to stand on your own.

SECTION 2

TOOLS, MATERIALS & BASIC TECHNIQUES

Why Tools Matter in Off-Grid Survival

Tools are the backbone of every project in this book. They save time, energy, and even your life.

But here's the truth most survival books won't admit:

You don't need a massive toolkit to survive off-grid.

You need the right tools—and you need to understand how to use them.

This section breaks down:

- Essential tools
- Budget alternatives
- What each tool is actually used for
- Basic techniques beginners must master

Everything is designed so you can start building immediately.

Core Survival Tools (The Must-Have Kit)

These are the tools you should have before attempting any major project.

1. Fixed-Blade Survival Knife

Purpose:

Cutting, carving, slicing, scraping, splitting, food prep, shaping wood.

Why It Matters Off-Grid:

A good knife replaces half your toolbox. It's your most reliable survival companion.

How to Use in Emergencies:

- Feather sticks for fire
- Carving tent stakes
- Cutting rope
- Preparing game or fish
- Scraping bark for tinder

Budget Alternative:

A stainless-steel kitchen knife + duct tape + a homemade sheath.

2. Folding Saw

Purpose:

Cutting small logs, trimming branches, shaping lumber for projects.

Why It Matters:

It's safer and more efficient than swinging an axe for beginners.

Emergency Use:

- Building shelters
- Cutting firewood
- Making frames for traps or structures

Budget Alternative:

Hacksaw with extra blades.

3. Hatchet or Small Axe

Purpose:

Chopping wood, splitting logs, shaping beams.

Why It Matters:

You need it for fire, structures, and repairs.

Emergency Use:

- Breaking into frozen ground
- Clearing debris
- Cutting emergency escape paths

Budget Alternative:

Heavy-duty machete.

4. Multi-Tool

Purpose:

Quick repairs, adjustments, small cuts, tightening bolts, removing screws.

Why It Matters:

Saves time and prevents small problems from turning into big ones.

Features to Look For:

Pliers, screwdriver heads, small blade, file, scissors.

Budget Alternative:

Basic plier + old screwdriver set + utility knife.

5. Fire-Starters

Types:

- Ferro rod
- Lighter
- Waterproof matches
- Magnesium block

Why It Matters:

Without fire, you lose heat, clean water, and cooked food.

Emergency Use:

- Starting fires in rain
- Generating sparks for improvised tinder
- Lighting stoves

Budget Alternative:

Old batteries + steel wool (instant flame).

6. Paracord & Rope

Purpose:

Shelter building, traps, backpacks, tying loads, securing structures.

Why It Matters:

Strength + flexibility + thousands of uses.

Emergency Use:

- Building shelters
- Making tools
- Hanging food storage
- Creating drying lines

Budget Alternative:

Strong twine, leftover electrical cables, or plant fibers.

7. Shovel (Collapsible or Full-Size)

Purpose:

Digging fire pits, trenches, latrines, planting, clearing debris.

Why It Matters:

There's no substitute for a good shovel during an emergency.

Emergency Use:

- Digging drainage around shelter
- Making earth ovens
- Burying waste

Budget Alternative:

Flat metal sheet + wooden handle.

8. Flashlight / Headlamp

Purpose:

Night operations, repairs, navigation.

Why It Matters:

Darkness multiplies danger. Light reduces it.

Emergency Use:

- Signaling
- Night cooking
- Repairing tools
- Navigating wooded area

Budget Alternatives:

Hand-crank flashlight or solar-powered lantern.

9. First-Aid Kit

Purpose:

Treating injuries before they become fatal.

Why It Matters:

In a grid-down world, even small wounds can become life-threatening.

Must-Haves:

- Bandages
- Antiseptic
- Pain relievers
- Tweezers
- Gloves
- Antibiotic ointment

Budget Alternative:

Homemade kit using household medical supplies.

10. Containers & Buckets

Purpose:

Storage, carrying water, mixing materials, food preservation.

Why It Matters:

Off-grid life involves constant movement of materials.

Budget Alternative:

Recycled paint buckets, old tins, jerry cans.

Materials You Should Always Stock

These materials appear repeatedly in survival projects. Think of them as the "building blocks" of off-grid engineering.

1. Wood (Different Types)

Softwood for kindling, hardwood for structures.

2. Metal Sheets

Used for roofs, stove bodies, heat shields, reflectors.

3. PVC Pipes

Water systems, traps, frames, airflow channels.

4. Clay / Soil

Earth ovens, insulation, filtration.

5. Stones & Gravel

Filtration, heat retention, foundation building.

6. Old Clothing / Fabric

Wicks, filters, carrying pouches.

7. Plastic Bottles & Containers

Water systems, solar projects, mini greenhouses.

8. Wire (Thick & Thin)

Binding, traps, tool handles, electrical repairs.

9. Duct Tape & Electrical Tape

Fix anything. Literally anything.

10. Scrap Metal

For blades, stove parts, brackets, hinges, hooks.

Basic Techniques Every Beginner Must Master (Crucial Skill Pages)

These techniques will appear in later projects. Learn them now.

Technique 1: Safe Knife Handling

- Always cut away from your body
- Use your non-dominant hand behind the blade
- Keep the knife sharp, it's safer
- Use controlled, small movements

Practice carving simple shapes before heavy tasks.

Technique 2: Building a Stable Frame

Most survival structures rely on a **triangle or square frame.**

Basic rules:

- Use thick branches as supports
- Lash corners tightly
- Test stability BEFORE adding weight
- Reinforce joints with rope or vines

Technique 3: Creating Natural Tinder

Good tinder = reliable fire.

Best options:

- Dry bark shavings
- Cotton cloth
- Dead grass
- Wood shavings
- Char cloth
- Scraped magnesium

Always gather more than you think you need.

Technique 4: Making Strong Lashings

Use paracord or plant fibers.

Common knots you must know:

- Square knot
- Clove hitch
- Bowline
- Taut-line hitch

These knots create the spine of your shelters and tools.

Technique 5: Heat Control for Cooking

Use:

- Hot coals for slow cooking
- Open flame for boiling
- Rocks to retain and distribute heat
- Clay and mud for insulation

Mastering heat helps prevent waste and improves fuel efficiency.

Technique 6: Water Filtration Basics

Improvised filters require:

- Stones (large)

- Gravel (medium)

- Sand (fine)

- Charcoal (activated if possible)

- Cloth for pre-filtering

Layering matters.

Flow rate matters.

Surface area matters.

Technique 7: Safe Wood Splitting

Don't swing wildly.

Use a stable surface.

Strike straight.

Use smaller pieces to build fires.

Safety first—injuries ruin survival chances.

Technique 8: Leveling & Ground Preparation

For structures, stoves, traps, or water systems, ground leveling is key.

Steps:

1. Clear debris

2. Flatten the area

3. Compact the soil

4. Mark boundaries

5. Ensure proper drainage

A stable base prevents collapse.

Minimum Budget Starter Kit (For Beginners)

This is the cheapest possible list you can build and still complete most projects in this book.

- Kitchen knife + file for sharpening
- Recycled bottles
- Old cloth & denims
- Tin cans
- Paracord (or strong ropes)
- Matches & lighter
- Basic first-aid supplies
- Buckets or gallons
- Hand saw
- Machete
- Flashlight
- Nails & wire
- Scrap wood
- Stones and gravel

With these alone, you can build shelters, filters, stoves, traps, ovens, reflectors, heating systems, and more.

Safety Principles for All Builds

Survival is useless if you get injured.

Follow these:

- Never cut toward yourself

- Keep your tools sharp

- Wear gloves when moving wood or metal

- Test structures before putting weight

- Keep fire a safe distance from shelters

- Ensure proper ventilation when using stoves

- Store water in safe, cleaned containers

- Don't inhale chemical fumes

- Don't use wet wood for structures

- Keep children away from builds

- Do not leave fires unattended

SECTION 3

72-HOUR EMERGENCY PLAN

A step-by-step guide for the first three days after the grid goes down.

Why the First 72 Hours Matter

The first three days of any crisis are the most chaotic. Power failures, communication loss, disrupted supplies, weather exposure, and confusion set in fast. What you do, and don't do in these hours determines whether you stay safe or spiral.

This plan removes the panic and gives you a clear, hour-by-hour roadmap you can follow with confidence.

HOUR 0–6: SECURE YOUR ESSENTIALS

1. Assess Immediate Safety

- Check for injuries, treat cuts, burns, or heavy bleeding fast.
- Move away from unstable structures, falling trees, flooded areas, or fire hazards.
- Turn off gas lines if you suspect leaks.

2. Secure Water (Top Priority)

You need at least 3–4 liters per person per day.

Do this immediately:
- Fill all containers, buckets, pots, bathtubs.
- Set up emergency filtration (DIY charcoal filter or commercial filter).
- If you have bleach, boil or disinfect early before contamination risk increases.

3. Start Your First Fire

Heat = safety + water purification + morale.

Choose a method:
- Ferro rod / magnesium

- Battery + steel wool
- Glass + sunlight
- Basic tinder bundle (cotton, dry leaves, shredded bark)

Once lit, stabilize the fire with a windbreak.

HOUR 6–24: SHELTER, SIGNALING & COMMUNICATION

1. Build or Strengthen Shelter

Your shelter must:
- Stay dry
- Block wind
- Retain heat
- Provide ventilation

Options:
- Lean-to with waterproof cover
- A-frame shelter
- Car survival shelter
- Reinforced indoor room (if safe)
- Insulate the inside using leaves, clothing, blankets, or cardboard.

2. Establish Communication

Grid down ≠ silence; you still need information.

Do the following:
- Test radio (AM/FM, shortwave, or hand-crank).
- Listen for emergency broadcasts.
- Set up bright-colored markers or mirrors for signaling.
- Create a check-in schedule with family/group.

3. Inventory Your Supplies

This is where most people make emotional, panicked mistakes.

Sort into:
- Water
- Food
- Fire supplies
- Tools
- First aid
- Light sources
- Power sources
- Defense/awareness tools

Set strict ration rules early.

DAY 2 (Hours 24–48): FOOD, POWER & RESOURCE MANAGEMENT

1.Food Management

Avoid eating too early, you burn energy digesting.

Do this instead:
- Identify high-calorie items first.
- Separate perishable and non-perishable foods.
- Begin using perishables immediately (fresh produce, bread).
- Keep canned/dry goods for later.

If foraging:
- Stick to known, safe plants and insects.
- Avoid mushrooms unless you're an expert.

2. Power Generation

Start simple. Build up.

Options:
- Solar panels + battery bank
- DIY hand-crank generator
- Bicycle generator
- Small portable power bank rotation
- Solar battery chargers

Focus on powering:
- Lights
- Phones (for notes, maps, offline tools)
- Radios

Avoid draining power on anything unnecessary.

3. Secure Your Perimeter

- You don't need to be paranoid, just practical.
- Set noise traps using cans, stones, or sticks.
- Keep a controlled fire burning for visibility and predators.
- Organize tools where you can reach them fast.
- Mark boundaries clearly.

DAY 3 (Hours 48–72): STABILIZATION & LONG-TERM PREP

This is where panic usually drops and strategy begins.

1. Stabilize Water & Food Supplies

- Set up a long-term water system (rain catchment or continuous filtration).
- Try fishing, trapping, or gathering if safe.
- Start preserving food: drying, smoking, or salting.

2. Improve Shelter Comfort & Safety

- Reinforce structural weaknesses.
- Add insulation and drainage channels.
- Raise your sleeping area off the ground.

Comfort reduces stress, and stress kills faster than hunger.

3. Build a Routine

Survival becomes easier when it's predictable.

Create daily cycles:
- Morning: water + fire maintenance
- Midday: repairs + scouting + resource gathering
- Evening: communication + security check + rest

4. Mental Stability

The third day is when exhaustion hits hard.

Do this:
- Stay warm
- Stay hydrated
- Keep your mind occupied (tasks, journal, planning)
- Avoid unnecessary risks

If you reach the end of Day 3 with your shelter, fire, water, and basic supplies intact, you're no longer in crisis mode. You're stabilizing.

SECTION 4

THE MAJOR SURVIVAL PROJECTS

PROJECT 1

BUILDING A BASIC OFF-GRID WATER FILTER
(Multi-Layer Sand & Charcoal System)

SUMMARY

Clean water is the most urgent need in any grid-down situation.
This project teaches you how to build a **long-lasting, effective, DIY water filtration unit** using simple materials like sand, gravel, charcoal, and a container.
It gives you water that's clean enough for drinking *after boiling*.

MATERIALS NEEDED

- 1 large plastic bottle, bucket, or PVC pipe

- Clean sand (fine)

- Gravel or small stones

- Activated charcoal (or homemade charcoal)

- Cloth, cotton, or coffee filter

- Knife or heated metal for cutting

- A container to catch filtered water

- Optional: mesh screen or thin wire

STEP-BY-STEP BUILD GUIDE

Step 1: Prepare the Container

Cut the bottom off the bottle or pipe, this becomes the top of your filter.
If using a bucket, punch a small hole in the base to allow slow dripping.

Step 2: Add the Cloth Filter Layer

Place a cloth at the very bottom (inside). This prevents sand from leaking out and acts as the first filter.

Step 3: Add the Charcoal Layer

Crush charcoal into small pieces (not powder).
Fill 2–3 inches of charcoal.
This removes odors, bacteria, and impurities.

Step 4: Add Fine Sand Layer

Add 3–4 inches of clean, washed sand.
Sand removes dirt, sediment, and fine particles.

Step 5: Add Gravel or Stones

Top the sand with 2–3 inches of gravel.
This helps distribute the water evenly and prevents sand disturbance.

Step 6: Add a Cloth or Mesh Top Layer

Prevents debris from falling inside when pouring water.

Step 7: Run Cleaning Water Through Once

Pour water through to wash the filter.
Discard the first batch.

Step 8: Start Filtering

Pour water slowly.
It will drip cleanly from the bottom.
Always boil after filtering.

SAFETY NOTES

- This filter removes particles and dirt, NOT viruses.
- Boil the filtered water for **5–10 minutes** before drinking.
- Replace sand and charcoal every 2–4 weeks if used daily.
- Avoid riverbeds with chemical pollution.

TROUBLESHOOTING

Water flows too fast:
– Add more sand or charcoal
– Reduce the size of the outlet hole

Water flows too slow:
– Stir the top gravel
– Reduce sand thickness
– Ensure cloth isn't blocking the exit

Water has smell:
– Replace charcoal
– Check source water

VARIATIONS

Budget Version

Use two plastic bottles, one inverted into the other, and charcoal from a fireplace.

Advanced Version

Use layered PVC pipes with a valve to control flow speed.

Off-Grid Permanent Setup

Create a 4-foot tall barrel filter with sand, charcoal, and gravel for a whole household.

BEST USE SCENARIOS

- Long-term grid-down events
- Bushcraft camps
- Emergency home systems
- Backup water when taps fail
- Rainwater purification

PROJECT 2

DIY ROCKET STOVE
(High-Efficiency Off-Grid Cooking Stove)

SUMMARY

A rocket stove is one of the most efficient off-grid stoves ever invented.

It burns small sticks and scraps of wood, produces intense heat, cooks fast, and uses very little fuel.

You'll build a simple, durable rocket stove using metal cans, bricks, or clay, depending on what you have available.

Great for:

- Cooking without electricity

- Boiling water

- Heating small spaces

- Outdoor emergencies

- Backyard preparedness

MATERIALS NEEDED

(Choose one style: Metal, Brick, or Clay)

Option A — Metal Can Rocket Stove

- 1 large tin can (paint can size)

- 1 medium tin can

- 1 small tin can or metal pipe (wood feed chute)

- Sand or ash (for insulation)
- Knife, tin snips, or metal cutter

Option B — Brick Rocket Stove
- 16–20 firebricks or cement bricks
- Flat ground or metal sheet to build on

Option C — Clay Rocket Stove
- Clay or mud
- Stones (small and medium)
- Metal pipe (optional)

Universal Materials
- Dry sticks or wood scraps
- Gloves
- Marker or chalk

STEP-BY-STEP BUILD GUIDE

Option A — METAL CAN ROCKET STOVE

Step 1: Prepare the Body
Cut a hole near the base of the large can.
This hole must fit your medium can.

Step 2: Insert the Burn Chamber
Slide the medium can through the hole.
This becomes the horizontal wood feed.

Step 3: Add the Vertical Chimney
Place the small can or metal pipe vertically inside the large can, connecting with the medium can.
This creates an L-shaped airflow system.

Step 4: Pack Insulation
Fill the surrounding space with sand, ash, or crushed stone to retain heat.

Step 5: Add Pot Support
Cut or hammer small metal tabs at the top so pots can sit over the chimney.

Step 6: Test Burn
Light small sticks inside the horizontal chamber.
You should see a strong upward flame.

Option B — BRICK ROCKET STOVE

Step 1: Create an L-Shape Channel
Stack bricks to form a sideways "L":
- Horizontal channel for wood feed
- Vertical channel for chimney heat rise

Step 2: Build the Walls
Add additional rows of bricks around the L-shape to form a tall chimney.

Step 3: Adjust Openings
Leave a small opening at the bottom of the vertical chimney for airflow.

Step 4: Stabilize the Build
Ensure bricks are tightly packed and level.

Step 5: Light and Test
Insert small dry sticks.
When fire burns correctly, you'll hear the signature "rocket" sound.

Option C — CLAY ROCKET STOVE

Step 1: Mix Clay or Mud
Combine clay and grass fibers for strength.

Step 2: Form a Thick Base

Build the L-shaped pathway using stones as reinforcement.

Step 3: Build Walls

Add clay all around to create a chimney. Leave a horizontal gap for wood insertion.

Step 4: Smooth and Shape

Pack clay tightly and shape the chimney top.

Step 5: Let it Dry

Air-dry for 24–48 hours before first use.

SAFETY NOTES

- Never use galvanized metal (it releases toxic fumes when heated).
- Keep the stove stable on flat ground.
- Do not use inside closed rooms without ventilation.
- Always burn dry wood to prevent smoke buildup.
- Metal cans get extremely hot, use gloves.

TROUBLESHOOTING

Low heat output:

- Wood is wet
- Chimney is too short
- Airflow is blocked
- Interior filled with ash (clean it)

Smoke coming from feed tube:

- Horizontal tube is too long
- Not enough vertical height

- Poor airflow
- Add small sticks instead of large logs

Stove collapses (brick/clay models):

- Ground is uneven
- Bricks not tightly stacked
- Clay walls too thin

VARIATIONS

Budget Version

Use 8–10 bricks only, simplest L-shape stove that still works.

Advanced Version

Add a double-wall system for hotter, cleaner burning.

Family-Use Version

Build a two-burner rocket stove side-by-side.

Bushcraft Version

Use stones and clay to make a temporary camp rocket stove.

BEST USE SCENARIOS

- Cooking during power outages
- Emergency boiling of water
- Outdoor cooking
- Camping
- Off-grid homesteads
- Urban survival setups
- Natural disaster aftermath

PROJECT 3

SHORT-TERM SURVIVAL SCENARIOS
(Urban, Rural & Wilderness)

Realistic, high-pressure situations designed to train instinct, decision-making, and fast adaptation.

Introduction to Short-Term Survival Scenarios

Short-term survival is all about what you do in the first **24 hours to 7 days** when things go wrong. No fancy gear. No comfort. Just your brain, your environment, and whatever scraps you can grab. This section drops you into scenarios that push your thinking, test your calm under stress, and force you to apply everything you've learned so far.

Each scenario includes:
- **Situation Overview**
- **Immediate Threats**
- **Your Priorities (S.T.A.R Method — Stop. Think. Assess. Respond.)**
- **Step-By-Step Survival Actions**
- **What Not to Do**
- **Gear That Would've Helped (Optional Learnings)**
- **Skill Focus**

Let's break it down scenario by scenario.

Situation Overview

A sudden nationwide blackout shuts down electricity, mobile networks, and banking. Streets are tense, stores are closing, and you're stuck in a crowded city with limited supplies.

Immediate Threats

- Looting, panic, aggressive crowds
- Loss of clean water access
- Spoiled food (fridges down)
- No communication
- No transport
- Night-time insecurity

S.T.A.R Priorities

- **Stop:** Don't panic or run outside blindly.
- **Think:** How long can you survive with what you have?
- **Assess:** Water? Food? Safe shelter?
- **Respond:** Secure resources, strengthen home defenses.

Step-By-Step Survival Actions

1. Secure Your Location

- Check doors/windows for weak points.
- Block balcony or open access points.
- Move valuables away from line of sight.

2. Do a Rapid Inventory Sweep

- Water (litres available)
- Food (calories available)
- Candles, lanterns, power banks
- First-aid kit
- Clothes for heat/cold

3. Conserve Resources

- Ration water immediately.
- Cook only perishables first.
- Keep lights minimal at night.

4. Create a Communication Plan

- Decide on meet-up points with family.
- Write important numbers on paper.

5. Stay Updated

- Use radio if available.
- Listen for emergency broadcasts.

6. Avoid Dangerous Zones

- Shops, ATMs, gas stations, or any place where crowds gather.

What Not to Do

- Don't wander outside "to see what's happening."
- Don't leave charging devices plugged overnight — fire risk.
- Don't confront looters.

Gear That Would've Helped

- Solar lantern
- Portable power bank
- Water purification tablets
- Foldable stove
- Radio

Skill Focus

Urban adaptation, security, rationing, emergency planning.

SCENARIO 2
ACCIDENTALLY STRANDED IN THE FOREST

Situation Overview

Your car breaks down deep in a forest road. Your phone battery dies. You have minimal food and no idea how far civilization is.

Immediate Threats

- Low visibility
- Predators
- Cold night temperatures
- Navigation errors
- Dehydration

S.T.A.R Priorities

- **Stop:** Don't walk blindly into the forest.
- **Think:** Staying with the vehicle increases rescue chance.
- **Assess:** Weather, time, resources.
- **Respond:** Create shelter, signal for help, get water.

Step-By-Step Survival Actions

1. Stay With the Vehicle

- It's the largest, most visible object you have.
- Rangers and rescuers search the road first.

2. Set Up a Shelter Using the Car

- Use interior as wind barrier.
- Combine car door + branches + jacket for a lean-to.

3. Find Water First

- Follow downhill slopes.
- Collect morning dew with cloth.
- Avoid stagnant pools.

4. Signal for Rescue

- Create a reflective signal using car mirrors.
- Make ground symbols (SOS or HELP) with branches/rocks.
- At night, use flash light (3 flashes = distress signal).

5. Make a Fire (If Safe)

- Dry leaves + bark as tinder.
- Car battery sparks (advanced technique).
- Position smoke so it's visible.

6. Food (Lowest Priority)

- You can survive 3–5 days without food.
- Focus on safety and water.

What Not to Do

- Don't walk more than 1 km from the car.
- Don't drink unpurified water unless dying of dehydration.
- Don't sleep on bare ground.

Gear That Would've Helped

- Ferro rod
- Emergency blanket
- Water filter straw
- Pocketknife

Skill Focus

Shelter building, signaling, fire-making, water sourcing.

SCENARIO 3
RURAL FLOODING & EVACUATION

Situation Overview

Heavy rainfall turns into flash flooding in a rural community. Roads are submerged. Homes are partially underwater. You need to move to higher ground.

Immediate Threats
- Fast water currents
- Hidden debris
- Electrocution from submerged power lines
- Hypothermia
- Contaminated water

S.T.A.R Priorities
- Stop: Don't enter flood water carelessly.
- Think: Which route has the highest ground?
- Assess: Can you evacuate safely?
- Respond: Move fast but smart.

Step-By-Step Survival Actions

1. Climb to Higher Ground Immediately
- Roof.
- Hill.
- Sturdiest tree.

2. Turn Off Power (If Safe)
- Prevent electrocution.

3. Pack a Fast Evac Kit (1 Minute)
- Water
- Dry clothes
- ID/documents
- Light source
- Phone + power bank
- First aid

4. Avoid Fast-Moving Water
- 15 cm of water can knock you over
- 60 cm can move a vehicle

5. Mark Your Location
- Bright cloth on roof
- Signal firefighters/boats

6. After Evacuation
- Purify all drinking water
- Avoid contact with contaminated flood debris
- Seek medical care for cuts (high infection risk)

What Not to Do
- Don't try to drive through flooded roads.
- Don't dive or swim in murky water.
- Don't walk barefoot—dangerous objects everywhere.

Gear That Would've Helped

- Waterproof bag
- Headlamp
- Life jacket
- Emergency whistle

Skill Focus

Flood navigation, fast decision-making, risk avoidance, emergency packing.

SCENARIO 4
DESERT STRANDED
(Extreme Heat Survival)

Situation Overview

Your bus breaks down in a semi-desert region. Shade is minimal. Water is extremely limited. Temperatures climb toward 40–45°C by noon.

Immediate Threats

- Rapid dehydration
- Heatstroke
- Sunburn
- Disorientation
- Night-time cold drops

S.T.A.R Priorities

- **Stop:** Don't walk under the sun aimlessly.
- **Think:** You have limited water; conserve energy.
- **Assess:** Shade sources, water supply, heat patterns.
- **Respond:** Create shade, ration water, travel at safe times.

Step-By-Step Survival Actions

1. Get Out of the Sun Immediately

- Use bus shade, clothing, tarp, or car doors.
- Create a low-profile shade shelter.

2. Water Management

- Sip water, don't gulp.
- Keep mouth closed to preserve moisture.
- Use cloth to reduce sweat evaporation.

3. Travel Only Early Morning or Late Evening

- Best window: 5–8 AM or 6–8 PM.
- Rest during midday heat.

4. Follow Tracks or Roads

- Never venture into open desert dunes.
- Roads = vehicles eventually pass.

5. Signal for Help

- Lay bright clothing on the ground.
- Use reflective surfaces for signaling aircraft.

6. Night Survival

- Desert cold can shock your body.
- Use clothing layers, dry brush, and windshield as a windbreaker.

What Not to Do

- Don't drink all your water at once.

- Don't walk at noon.
- Don't sit on bare sand (heat transfers fast).

Gear That Would've Helped
- Wide-brim hat
- Emergency blanket
- Electrolyte packets
- Solar-powered phone charger

Skill Focus
Heat management, water conservation, shelter improvisation.

SCENARIO 5
STRANDED AT SEA
(Small Boat or Capsized Kayak)

Situation Overview

Strong winds overturn your kayak/boat. You're drifting on open water with limited gear and uncertain rescue time.

Immediate Threats
- Dehydration
- Saltwater ingestion
- Hypothermia (even in hot climates)
- Sun exposure
- Sharks/large marine animals

S.T.A.R Priorities
- **Stop:** Stay with your flotation device.
- **Think:** Your boat is your signal beacon.
- **Assess:** Weather, currents, injuries.
- **Respond:** Conserve energy, improve visibility.

Step-By-Step Survival Actions

1. Stay With the Boat or Flotation
- It's 10× more visible than a floating person.

2. Increase Visibility
- Wave bright clothing.
- Use reflective objects.
- Create rhythmic signals (whistle, tapping).

3. Protect Against Sun & Salt
- Cover exposed skin.
- Keep eyes away from salt spray.
- Never drink seawater.

4. Collect Water
- Use rainwater catchment (clothing, containers, plastic).
- Condensation trick with plastic bag + plant material (if near land).

5. Avoid Heat Loss
- Curl into Heat Escape Position.
- Keep legs together, arms crossed.

6. Food
- Low priority.
- You can survive 3–7 days with zero food.

What Not to Do
- Don't swim toward "distant land", optical illusions are common.
- Don't hang limbs unnecessarily, attracts marine predators.
- Don't drink rainwater that collected salt spray.

Gear That Would've Helped

- Life jacket
- Signal mirror
- Distress flare
- Waterproof dry bag

Skill Focus

Energy conservation, water sourcing, signaling, calm endurance.

SCENARIO 6
WILDLIFE ENCOUNTER
(Bear, Snake or Big Cat)

Situation Overview

While hiking, you accidentally enter the territory of a large wild animal. It sees you, and there's no clear exit route.

Immediate Threats

- Aggressive charge
- Venomous bites
- Territorial attack
- Panic-driven mistakes

S.T.A.R Priorities

- **Stop:** Freeze. Quick movement triggers attacks.
- **Think:** Identify the species.
- **Assess:** Distance, escape path, your posture.
- **Respond:** Choose the right behavioral response.

Step-By-Step Survival Actions

If it's a big cat (lion, leopard, cougar):

- Make yourself look big.
- Maintain eye contact.
- Back away slowly.
- Never turn your back or run.

If it's a bear:

- For black bears: Stand tall. Make noise. Throw objects.
- For brown/grizzly: Don't scream. Don't challenge.
- Back away very slowly.
- If attacked: Play dead.

If it's a snake:

- Freeze.
- Step back slowly.
- Don't stomp or throw objects.
- If bitten: Stay still, keep bite below heart, don't cut/suck.

What Not to Do

- Don't run. Almost every predator is faster than you.
- Don't scream in panic.
- Don't make sudden movements.

Gear That Would've Helped

- Bear spray
- Hiking pole
- First-aid kit
- Thick boots

Skill Focus

Calm decision-making, animal behavior, body language control.

SCENARIO 7
HOUSE FIRE ESCAPE
(Night-Time Emergency)

Situation Overview

You wake up at 2 AM to smoke entering your room. Electricity is out. Heat is increasing rapidly, and the fire seems close.

Immediate Threats

- Smoke inhalation
- Toxic fumes
- Low visibility
- Collapsing structures
- Being trapped

S.T.A.R Priorities

- **Stop:** Don't run into dark hallways blindly.
- **Think:** Identify the closest exit.
- **Assess:** Fire direction, smoke levels, obstacles.
- **Respond:** Stay low, move fast, escape smart.

Step-By-Step Survival Actions

1. Stay Low
- Smoke rises. Crawl under it.

2. Check Doors Before Opening
- Use back of hand, if hot, don't open.
- If safe, open slowly and stay low.

3. Use Clothing as Smoke Filters
- Wet cloth over mouth and nose if possible.

4. Know Your Escape Path
- Avoid elevators.
- Move toward outdoor exits or windows.

5. If Trapped
- Seal gaps under door with cloth.
- Signal from window using a bright cloth or flashlight.
- Stay visible.

6. Once Outside
- Do not re-enter the building.
- Call emergency services immediately.

What Not to Do
- Don't open windows near the fire, oxygen increases flames.
- Don't stand up straight inside smoke-filled rooms.
- Don't try to "grab valuables."

Gear That Would've Helped
- Smoke hood
- Fire extinguisher
- Fire blanket
- Emergency escape ladder

Skill Focus

Fast evacuation, smoke navigation, emergency response.

PROJECT 4

DIY SAND + GRAVEL WATER FILTRATION SYSTEM (Multi-Layer Survival Water Filter

SUMMARY

A sand–gravel filtration system is one of the most reliable off-grid water filters ever created. It removes:

- Dirt
- Sediment
- Micro-organisms
- Odor
- Suspended particles
- Cloudiness

It's not a purifier by itself, but it turns dirty river or rainwater into clear, drinkable water once combined with boiling or solar disinfection.

This project teaches you to build a powerful, multi-layer filtration column using materials you can find anywhere.

Ideal for:

- Homesteads
- Emergency camps
- Flood aftermath
- Rural survival
- Long-term off-grid setups

MATERIALS NEEDED

Core Materials

- 1 large container (bucket, drum, or big plastic bottle)
- Clean fine sand
- Coarse sand
- Small gravel
- Large gravel/rocks
- Cotton cloth or coffee filters
- Activated charcoal (optional but highly recommended)
- Knife or drill (for cutting hole)
- Clean container for collecting filtered water
- Plastic tap (optional upgrade)

Optional But Powerful Add-ons

- **Mesh screen**
- **Charcoal from wood fire (if activated charcoal is unavailable)**
- **UV-resistant bucket (long-term durability)**
- **Second bucket for settling tank**

BUILDING THE SYSTEM (STEP-BY-STEP)

We're building a **5-Layer Gravity Filtration Column.**

STEP 1 — Prepare the Container

- Use a bucket or large plastic container.
- Cut a small hole at the bottom (just big enough for water to drip out).
- Add a tap if you want a cleaner flow.

Goal: Controlled slow filtration.

STEP 2 — Install the First Protection Layer (Cloth Filter)

- Place cotton cloth or coffee filter inside at the bottom.
- This prevents sand from leaking through the exit hole.

Goal: Keep the filter clean and structured.

STEP 3 — Add Layer 1: Large Gravel

- Fill the bottom with 2–3 inches of large stones.
- This creates drainage and prevents clogging.

What It Removes: Big debris like leaves, sticks, mud chunks.

STEP 4 — Add Layer 2: Small Gravel

- Add a 2-inch layer of small gravel.
- This forms a tighter barrier and evens out flow.

What It Removes: Finer debris and sand-like particles.

STEP 5 — Add Layer 3: Coarse Sand

- Add 3–4 inches of coarse sand.
- Wash the sand before adding to remove dirt.

What It Removes: Suspended solids, turbidity, cloudy-looking particles.

STEP 6 — Add Layer 4: Fine Sand

- Add another 4–5 inches of well-washed fine sand.
- This is your main filtration layer.

What It Removes:

- Small particles
- Micro-organisms
- Odor
- Micro-sediments
- Fine impurities

STEP 7 — Add Layer 5: Charcoal (Optional but HIGHLY recommended)

- Activated charcoal gives the cleanest taste.
- If using homemade charcoal, crush into small pieces and wash quickly to remove ash.

What It Removes:

- Chemicals
- Bad taste
- Odor

- Pesticide residues
- Organic pollutants

STEP 8 — Add a Top Cloth Layer
- A cloth over the top prevents dirt from falling into the sand.

STEP 9 — First Flush
Run 2–3 liters of water through the system and throw that away.
This clears dust and settles the layers.

STEP 10 — Filtration Begins
Pour dirty water on top.
Within minutes, clean, clear water drips from the bottom.

HOW TO USE THE FILTER
For Maximum Safety:
After filtration, you must either:
- **Boil the water for at least 5 minutes,**
OR
- **Use solar disinfection (SODIS):**
    o Put filtered water in clear bottles
    o Leave in direct sunlight for 6 hours
This kills bacteria and viruses completely.

MAINTENANCE
Daily Use
- Clean the top cloth.
- Scrape and remove the top 1 cm of fine sand every 3–5 days (it collects contaminants).

Monthly (Heavy Use)
- Remove all layers
- Wash thoroughly
- Let dry

- Rebuild system
- Replace charcoal

Lifespan

- Sand and gravel: virtually unlimited
- Charcoal: replace every 2–3 months

COMMON ISSUES & FIXES

Water Filtering Too Slowly

- Sand too fine
- Layers compacted
- Clogged top cloth

Fix: Stir top sand layer lightly and clean cloth.

Water Comes Out Cloudy

- Sand wasn't washed before use

Fix: Rinse sand until water runs clear.

Water Has Bad Taste

- Charcoal missing or expired

Fix: Add fresh charcoal layer.

SAFETY WARNINGS

- Never collect water near industrial areas or downstream of chemical waste.
- Avoid stagnant water if possible.
- Always disinfect filtered water before drinking.
- Don't touch the clean-water outlet with dirty hands or containers.
- Keep system covered to avoid insects or animals entering.

VARIATIONS

Portable Bottle Filter Version

Use a large soda bottle with improvised layers.

Bucket-To-Bucket Slow Sand Filter

One bucket filters into another.

Allows high-volume water production.

High-Pressure Ceramic Upgrade

Add ceramic filter at bottom for advanced purification.

Two-Stage Extended System

Bucket #1: Settling tank

Bucket #2: Sand filter

Bucket #3: Charcoal purifier

This mimics professional off-grid systems.

BEST USE SCENARIOS

- Emergency flooding
- Rural communities
- Camping trips
- Bushcraft shelters
- Homesteads
- Disaster aftermath
- Long-term water supply for off-grid families

PROJECT 5

DIY HAND-CRANK GENERATOR (Emergency Off-Grid Power Source)

SUMMARY

This project shows you how to build a functional **hand-crank generator** that produces electricity using nothing but human power.

In survival situations, this little machine becomes a lifeline, giving you power when the grid is dead.

Your generator can power:

- Phones
- Radios
- LED lights
- Rechargeable batteries
- GPS units
- Small USB-powered devices

It's reliable, simple, and requires no fuel — perfect for long-term off-grid living.

MATERIALS NEEDED

Core Components

- DC motor (12V or 24V preferred)
- Hand crank (metal rod, old bicycle pedal, drill handle, etc.)
- Diode (to prevent reverse flow)
- 1 USB charging module (5V buck converter)
- Small battery (optional but useful)
- Wires (insulated)

- Soldering iron (or twist wire method if no tools)
- Mounting board (wood or metal)
- Screws, zip ties, or brackets

Optional Upgrades
- Voltage regulator
- On/off switch
- Flywheel (makes cranking smoother)
- Multimeter
- Rechargeable battery pack (18650 cells)

HOW IT WORKS (Simple Breakdown)

A DC motor spins → produces voltage.

The faster you spin it → the more voltage you make.

The crank is your "engine."

The USB module steps it down to 5V for safe charging.

A 12V motor can easily generate 5–18 volts by hand.

Perfect for emergency power.

BUILD STEPS

STEP 1 — Prepare the Motor
- Clean the shaft.
- If shaft is smooth, wrap with tape or attach small metal clamp.

This gives the crank something to grip.

STEP 2 — Attach the Crank

You can use:
- Old bicycle pedal
- Short metal rod
- Wooden handle
- Broken drill handle

Secure it tightly onto the motor shaft.

Goal: Smooth rotational force without slipping.

STEP 3 — Mount the Motor

- Screw the motor onto a flat board.
- Ensure the crank can spin freely.
- Add side supports if crank is heavy.

Why?

A moving motor is inefficient and unsafe.

STEP 4 — Install the Diode

- Connect diode to the positive wire.
- Arrow direction should point towards your output.

Purpose:

Prevents electricity from flowing backward and damaging the motor.

STEP 5 — Connect the USB Converter

- Take the two wires from the motor (positive & negative).
- Connect them to the input side of your USB buck module.
- Solder or twist wires tightly.

This step ensures your generator always outputs a safe 5V USB power.

STEP 6 — OPTIONAL: Add a Battery Buffer

Connect a small rechargeable battery pack in between the generator and USB output.

Advantages:

- Crank once, power later
- Smooth power delivery
- Store energy at night or in storms

STEP 7 — Test the Output

Start cranking:

- Slow crank = 3–6V
- Medium crank = 6–10V
- Fast crank = 12–18V

Your USB module should stabilize it to exactly 5V.

Use a multimeter if you have one.

STEP 8 — Final Assembly & Reinforcement
- **Tie wires neatly**
- **Screw down loose parts**
- **Add handle grip**
- **Test with a small LED first**
- **Then test with phone or radio**

Your hand-crank generator is now field-ready.

HOW MUCH POWER CAN YOU EXPECT?

Light Devices
- LED lamps → easy
- Radios → easy
- Emergency beacons → easy

Medium Devices
- Phones → very doable
- USB power banks → doable with patience

Heavy Devices
- Laptops → you'll need a stronger build
- Car batteries → not recommended

Average output:
- 3–15 watts depending on speed and motor quality.

MAINTENANCE
- Keep motor shaft lubricated.
- Avoid over-cranking.
- Keep generator dry and dust-free.
- Replace worn wires.
- Test battery pack weekly (if installed).

TROUBLESHOOTING

Weak power output

- Cranking too slow
- USB buck converter faulty
- Wires loose
- Motor worn out
- No diode installed (voltage drops backward)

Phone not charging

- Voltage too low
- USB converter overheating
- Use faster cranking speed
- Try another cable

Crank feels stiff

- Shaft misaligned
- Over-tightened screws
- Add lubrication
- Reduce gear friction

UPGRADES (FOR ADVANCED USERS)

1. Add a Flywheel

- Makes cranking smoother
- Stores rotational energy

2. Gear Ratio System

- Faster motor spin with same hand effort

3. Multi-Device Output

- Add two USB ports
- Add indicator LED
- Add over-voltage protection

4. Combine With Solar

- Charge during the day with solar
- Crank at night during emergencies

5. Waterproof Casing

- For outdoor or rainy-season use

BEST USE SCENARIOS

- Off-grid homesteads
- Emergency kits
- Camping trips
- Natural disasters
- Survival shelters
- Car breakdowns
- Flood & blackout rescue
- Bushcraft field power
- War/conflict areas
- Remote villages
- Daily household backup

PROJECT 6

DIY SOLAR DEHYDRATOR (Preserve Food Off-Grid)

SUMMARY

A solar dehydrator uses sunlight to remove moisture from fruits, vegetables, and meat, extending shelf life **without electricity**.

It's perfect for:

- Long-term storage
- Emergency preparedness
- Bushcraft and camping
- Off-grid homesteads
- Reducing food spoilage

This project allows you to preserve:

- Fruits (mango, banana, apple, pineapple)
- Vegetables (tomato, carrot, peppers)
- Meats (jerky, fish)
- Herbs (basil, thyme, mint)

The dehydrator prevents bacterial growth by lowering moisture content while retaining nutrients and flavor.

MATERIALS NEEDED

Core Materials

- Wooden or cardboard frame
- Clear acrylic sheet, plexiglass, or transparent plastic

- Mesh trays or perforated racks
- Screws, nails, or glue
- Hinges (optional, for access door)
- Black paint (to absorb heat)
- Wire mesh or netting (keep insects out)
- Reflective surfaces (aluminum foil optional, for increased heat)

Optional Upgrades
- Small thermometer
- Adjustable vent
- Solar-powered fan
- Foldable/portable design
- Adjustable angles for sun tracking

BUILDING THE SOLAR DEHYDRATOR

STEP 1 — Construct the Base Frame
- Build a rectangular frame 2-3 feet wide, 3-4 feet long, and 1-2 feet high.
- Use wood or heavy-duty cardboard reinforced with glue and tape.
- Paint the inside black to **absorb heat efficiently.**

Goal: Maximize internal temperature for dehydration.

STEP 2 — Add Trays/Racks
- Install **mesh trays** in layers (2-5 levels depending on size).
- Ensure air can flow under and around trays.
- Space trays about 2-3 inches apart.

Goal: Even air circulation for consistent drying.

STEP 3 — Attach the Transparent Top Cover
- Place clear acrylic or plastic sheet as the top cover.
- Angle the cover toward the sun (30-45° ideal) to catch maximum sunlight.
- Optional: add a hinge for easy tray access.

Goal: Greenhouse effect: trap heat while letting moisture escape.

STEP 4 — Ventilation

- Cut small vent holes at the top or sides.
- Cover vents with fine mesh to prevent insects.
- Optional: add adjustable vent covers to regulate airflow.

Goal: Moist air escapes while warm air stays trapped.

STEP 5 — Add Reflective Surfaces

- Place aluminum foil or mirrored surfaces outside the box to direct sunlight into the dehydrator.
- Optional: angle reflectors to maximize sun exposure throughout the day.

Goal: Boost internal temperature for faster drying.

STEP 6 — Test Temperature

- Ideal drying temperature: 50–60°C (120–140°F) for fruits/vegetables, 60–70°C (140–160°F) for meat.
- Use thermometer if available; otherwise, a hand test works — warm but not burning hot.

STEP 7 — Start Drying

1. Slice food evenly (0.25–0.5 inches thick for fruits, 0.25 inches for meat).
2. Place food in single layers on trays.
3. Cover with mesh/netting if needed.
4. Rotate trays every 2–3 hours for even drying.
5. Drying time:
   o Fruits: 8–12 hours
   o Vegetables: 6–10 hours
   o Meat: 10–14 hours

MAINTENANCE

- Clean trays after every use.
- Wipe internal surfaces monthly.

- Check plastic/acrylic for cracks.
- Replace mesh if torn.
- Store dehydrator in shade when not in use.

TROUBLESHOOTING

Food not drying
- Sunlight insufficient → adjust angle or add reflectors
- Ventilation blocked → clear vents
- Food slices too thick → slice thinner

Food molding after drying
- Moisture not fully removed → dry longer
- Storage containers not airtight → use sealed jars or ziplock bags
- Humid environment → add silica packets or dry food further

Dehydrator overheating
- Vent slightly more
- Use thermometer to monitor

VARIATIONS

Portable Version
- Foldable cardboard + mesh trays
- Sun exposure outdoors
- Great for camping and small-scale use

Advanced Version
- Solar-powered fan for continuous airflow
- Adjustable angles for sun tracking
- Multiple layers for high-volume drying

Off-Grid Permanent Version
- Wooden, black-painted box with fixed glass top
- Multi-layer racks
- Permanent placement in sunny area
- Can dry large amounts of fruits and vegetables for long-term storage

BEST USE SCENARIOS

- Emergency preparedness
- Off-grid living and homesteads
- Camping trips and bushcraft
- Post-disaster food preservation
- Seasonal harvest storage
- Survival pantry creation

PROJECT 7

DIY COMPOSTING TOILET (Off-Grid Sanitation Solution)

SUMMARY

A composting toilet allows you to manage human waste safely **without water or sewage systems**, turning it into usable compost over time.

Perfect for:

- Off-grid homes
- Emergency shelters
- Bushcraft camps
- Rural survival
- Disaster aftermath

It prevents:

- Disease spread (cholera, E. coli, etc.)
- Ground and water contamination
- Smells and pests

It also produces **fertilizer** for gardens once the composting process is complete.

MATERIALS NEEDED

Core Materials

- Large plastic or wooden bucket (5–10 gallons)
- Toilet seat or simple seat cutout
- Sawdust, wood shavings, or coconut coir (carbon source)
- Lid to cover bucket
- Small trowel or scoop
- Gloves

- Ventilation pipe (optional but recommended)
- Drill or knife

Optional Upgrades
- Second bucket for rotation (continuous use)
- Compost bin or secondary container
- Activated charcoal (for odor control)
- Screen or mesh for ventilation
- Sealable lid for transportable toilet

HOW IT WORKS
- Human waste + carbon source → stored in airtight/semi-aerobic container
- Microorganisms break down the waste
- Moisture controlled to prevent odor
- After 6–12 months, compost is safe to handle and can fertilize non-edible plants

Key principle: Nitrogen (waste) + Carbon (sawdust/peat) = safe composting

BUILDING THE COMPOSTING TOILET

STEP 1 — Prepare the Container
- Choose a sturdy bucket (plastic is easiest).
- Drill small holes near the top sides for minimal ventilation (optional).
- Ensure bucket can support the seat.

STEP 2 — Attach the Toilet Seat
- Use a ready-made toilet seat or cut a hole in a flat board to fit comfortably.
- Secure seat to prevent wobbling.

Goal: Comfort and stability during use.

STEP 3 — Prepare the Carbon Layer
- Add 2–3 inches of sawdust, wood shavings, or coir at the bottom of the bucket.
- This absorbs liquids and helps start decomposition.

Tip: Keep a small scoop nearby for easy use.

STEP 4 — Daily Use Procedure

1. After each use, cover waste completely with 1–2 inches of carbon source.
2. Stir lightly if needed to mix layers.
3. Keep the lid closed when not in use.
4. Maintain dryness — urine should be separated if possible (optional).

STEP 5 — Ventilation (Optional but Recommended)

- Attach a small pipe through the lid or wall for venting gases.
- Direct airflow away from living areas.
- Cover vent with mesh to prevent insects.

STEP 6 — Manage Moisture & Odor

- Add extra sawdust for wet waste.
- Do not let liquid pool — it slows composting and causes smell.
- Mix occasionally to aerate.

STEP 7 — Secondary Bucket Rotation (Optional)

- When the bucket is full, swap to a second bucket.
- Allow the first bucket to compost for 6–12 months.
- Compost is safe for use on non-edible plants.

MAINTENANCE

- Always keep lid closed.
- Clean seat regularly with minimal water and natural disinfectant.
- Replace sawdust supply as needed.
- Inspect for pests or leaks weekly.
- Rotate buckets to maintain continuous use.

TROUBLESHOOTING

Smell Issues

- Cause: Too wet or insufficient carbon
- Fix: Add more sawdust, stir, improve ventilation

Pest Problems

- Cause: Open lid, exposed waste
- Fix: Cover immediately, use mesh, keep lid closed

Slow Composting

- Cause: Low temperature, too wet, or too dry
- Fix: Maintain moisture ~50%, keep in warm location, stir occasionally

Overflowing Bucket

- Cause: Excess urine or improper carbon layer
- Fix: Separate urine if possible, increase carbon, swap bucket sooner

VARIATIONS

Portable Toilet Version

- Smaller bucket with sealable lid
- Great for camping or emergency kits

Permanent Off-Grid Home Version

- Wooden cabin-style box with removable buckets
- Ventilation pipe to outside
- Compost storage chamber underneath

High-Volume Community Version

- Multiple linked units
- Separate urine chamber
- Large composting pit for rotation

BEST USE SCENARIOS

- Off-grid homesteads
- Camping or bushcraft survival
- Disaster response shelters
- Remote rural areas
- Eco-friendly toilet solution

PROJECT 8

DIY SOLAR STILL (Emergency Water Purification)

SUMMARY

A solar still is a simple device that uses sunlight to **distill water** from contaminated sources like:

- Saltwater
- Muddy river water
- Plant moisture
- Urine (in extreme survival situations)

It works by **evaporation and condensation**, leaving impurities behind.

Perfect for:

- Desert survival
- Coastal survival
- Flood aftermath
- Long-term off-grid scenarios

A solar still can produce **1–2 liters of clean water per day** depending on sunlight and setup.

MATERIALS NEEDED

Core Materials

- Large container or pit (plastic bowl, metal tray, or dug hole)
- Clear plastic sheet (transparent, flexible)
- Small cup or jar (to collect distilled water)
- Rocks or weights
- Optional: tubing for condensed water

Optional Enhancements

- Reflective surfaces (aluminum foil)
- Black container or lining to absorb heat
- Shade cover for reduced evaporation loss
- Multiple stills for higher yield

BUILDING THE SOLAR STILL

STEP 1 — Select a Location

- Choose a sunny spot with minimal wind.
- If digging a pit, ensure soil is moist.
- Maximize sun exposure throughout the day.

STEP 2 — Dig a Pit (If Using Ground)

- Diameter: 3–4 feet
- Depth: 1–2 feet
- Place a small cup or jar in the center of the pit to collect water.

STEP 3 — Add Water Source or Moisture

- Pour saltwater, muddy water, or plant material around the jar (but not inside it).
- Optional: add wet leaves or urine (extreme survival) for moisture.
- Do not overflow — keep the cup dry.

STEP 4 — Cover With Plastic Sheet

- Stretch the transparent plastic over the pit/container.
- Seal edges with rocks, soil, or tape to prevent vapor loss.
- Place a small **rock in the center of the sheet** to create a low point above the collection cup.

Goal: Condensed water drips into the cup instead of running off.

STEP 5 — Solar Distillation Process

1. Sun heats the water/soil/plant material.

2. Water evaporates, leaving contaminants behind.
3. Vapor condenses on the underside of the plastic sheet.
4. Condensed droplets run down to the low point and drip into the cup.

STEP 6 — Collect Water
- Wait 6–12 hours depending on sunlight and weather.
- Remove plastic carefully to prevent contamination.
- Drink distilled water directly or store in a clean container.

MAINTENANCE & SAFETY
- Clean collection cup daily.
- Check plastic sheet for tears.
- Reposition the still to follow sunlight if possible.
- Always avoid touching condensed water with dirty hands.

Note: Distilled water may lack minerals; use only in emergencies or supplement with electrolyte sources if available.

TROUBLESHOOTING

Little to no water collected
- Cause: insufficient sunlight, poor sealing, dry soil
- Fix: move to sunnier spot, seal edges tightly, add more moisture

Water contaminated
- Cause: cup touched by dirty water
- Fix: ensure cup placement is above water line, avoid spills

Plastic sheet collapses
- Cause: wind or improper weighting
- Fix: use multiple rocks, tie edges, use sturdier sheet

VARIATIONS

1. Portable Bottle Still
- Cut the top of a large bottle, place a cup inside, cover with plastic wrap.
- Great for desert or camping situations.

2. Large-Scale Solar Still

- Use tarp over large pit with multiple collection cups.
- Can provide water for small groups.

3. Tubing System

- Condensate can be channeled through tubing to distant container for easier collection.

BEST USE SCENARIOS

- Desert survival or stranded in arid environments
- Coastal survival with seawater
- Flood or post-disaster water shortages
- Off-grid homesteads for backup water
- Survival training and bushcraft exercises

PROJECT 9

DIY EMERGENCY SHELTER
(Quick-Deploy Survival Shelter)

SUMMARY

An emergency shelter protects you from:

- Sun and heat
- Cold and wind
- Rain and storms
- Insects and wildlife

This project teaches you to build a **temporary yet effective shelter** using minimal materials. Ideal for:

- Wilderness survival
- Natural disasters
- Bushcraft and camping
- Off-grid emergencies

The shelter is lightweight, fast to assemble, and can be adapted to any environment.

MATERIALS NEEDED

Core Materials

- Tarp, poncho, or large plastic sheet
- Rope or paracord (10–20 meters)
- Stakes (wood, metal, or rocks)
- Knife or multi-tool
- Branches, poles, or sticks
- Optional: leaves, grass, or pine boughs for insulation

Optional Enhancements

- Lightweight emergency blanket or foil for insulation
- Mosquito net
- Hammock setup
- Reflective surface to trap heat
- Small stones for anchoring

SELECTING A LOCATION

1. Avoid hazards: Dead trees, flooding zones, animal trails.
2. Look for natural windbreaks: Rocks, trees, or hills.
3. Ground surface: Flat, dry, and free of debris.
4. Sun exposure: Favor morning sun for warmth; avoid scorching afternoon sun if hot climate.

TYPES OF EMERGENCY SHELTERS

1. Lean-To Shelter

- One of the fastest and easiest designs
- Provides wind and rain protection

Setup Steps:

1. Tie rope between two trees at ~1.5–2 meters high.
2. Drape tarp over rope like a slanted roof.
3. Secure edges to ground with stakes or rocks.
4. Add leaves or branches for insulation under the tarp.

Best For: Wind protection, quick rain cover.

2. A-Frame Shelter

- Durable, good for cold climates
- Offers both wind and rain protection

Setup Steps:

1. Tie rope between two supports.
2. Fold tarp over rope evenly.
3. Stake sides to ground, forming a tent-like "A" shape.

4. Add insulation (leaves, pine boughs) inside or under sleeping area.

Best For: Cold, wet, and forested environments.

3. Debris Hut (Natural Materials)

- Uses leaves, sticks, and boughs
- Excellent insulation in cold climates

Setup Steps:

1. Build frame with two sturdy sticks on the ground as base and one as ridgepole.
2. Lean branches against the ridgepole on both sides to form triangle.
3. Cover frame with leaves, grass, or pine boughs.
4. Layer thickly for insulation; 30–50 cm is ideal.

Best For: Cold weather, minimal tarp availability.

4. Tarp Hammock Shelter

- Off-ground sleeping; protects from insects and moisture
- Quick and portable

Setup Steps:

1. Tie tarp between two trees above ground (~1–1.5 meters high).
2. Drape tarp over suspended rope or hammock.
3. Stake or secure sides for wind protection.

Best For: Wet ground, tropical climates, insect-heavy areas.

SHELTER SETUP TIPS

- Angle tarp for rain runoff
- Keep entrance facing away from prevailing wind
- Elevate sleeping area if possible
- Use natural insulation for cold climates
- Keep interior dry by removing debris before sleeping

MAINTENANCE & SAFETY

- Inspect knots and stakes daily
- Remove debris or puddles from inside

- Check for insect nests or animal tracks
- Replace or reinforce tarp if torn
- Avoid building under dead trees or unstable ground

TROUBLESHOOTING

Shelter collapses
- Cause: Weak support poles, loose rope
- Fix: Reinforce frame, tighten knots, stake corners firmly

Water leaks inside
- Cause: Incorrect tarp angle, gaps at edges
- Fix: Adjust tarp, add extra layer, elevate sleeping surface

Too cold inside
- Cause: Poor insulation, wind gaps
- Fix: Add leaves, blankets, or reflective emergency sheet

Too hot inside
- Cause: Full sun exposure
- Fix: Ventilation at one end, shade with branches

VARIATIONS

Quick Forest Shelter
- Lean-to against fallen tree or rock face
- Rapid deployment in wooded areas

Snow Survival Shelter
- Use snow walls for insulation
- Compact snow for windbreaks
- Tunnel entrance to reduce heat loss

Urban Disaster Shelter
- Inside abandoned building: tarp partitions, debris insulation
- On rooftops: windproof tarp lean-to

BEST USE SCENARIOS

- Wilderness survival
- Natural disasters (earthquake, hurricane, flood)
- Bushcraft expeditions
- Camping in harsh weather
- Off-grid homestead emergencies
- Evacuation zones

PROJECT 10

DIY FIRE STARTER KIT (Emergency Fire Essentials)

SUMMARY

Fire is crucial for:

- Warmth and survival in cold conditions
- Cooking and boiling water
- Signaling for help
- Light and protection from wildlife
- Sterilizing tools and containers

A **DIY fire starter kit** ensures you can produce fire quickly under any circumstances, even when conventional matches or lighters fail.

This kit can be compact, portable, and reusable, making it ideal for:

- Bushcraft and camping
- Disaster preparedness
- Off-grid survival
- Emergency kits

MATERIALS NEEDED

Core Materials

- Ferrocerium rod (ferro rod) or magnesium block
- Striker or knife edge (for sparks)
- Cotton balls or dryer lint (tinder)
- Wax (optional, to make waterproof tinder)
- Small container or tin to hold tinder and ferro rod
- Sandpaper (optional, for magnesium shavings)

Optional Additions

- Flint and steel kit
- Char cloth
- Fireproof gloves
- Emergency candles
- Mini lighter or waterproof matches

BUILDING THE FIRE STARTER KIT

STEP 1 — Prepare the Tinder

- Take cotton balls or dryer lint.
- Optional: dip in melted wax and let cool for waterproof tinder.
- Store tinder in a small, airtight container or tin.

Goal: Quick ignition from spark.

STEP 2 — Prepare Ferro Rod

- Attach ferro rod to a handle or keep in small tin.
- Ensure rod is protected from moisture.
- Striker should be a sharp metal edge or back of knife.

Tip: Practice scraping the rod to produce sparks before going into the field.

STEP 3 — Prepare Magnesium Shavings (Optional for Extreme Wet Conditions)

- Scrape magnesium block into small pile.
- Mix with tinder in tin.
- Magnesium ignites quickly and burns very hot — useful for damp conditions.

STEP 4 — Assemble Kit

- Place tinder, ferro rod, striker, and optional magnesium shavings in compact tin.
- Keep tin dry and airtight.
- Add label or marking for quick identification in emergencies.

Optional: Add mini lighter for backup ignition.

HOW TO USE FIRE STARTER KIT

1. Prepare a small fire pit with kindling (twigs, dry leaves, bark).
2. Pull out cotton ball or waxed tinder.
3. Scrape ferro rod with striker to produce sparks directly onto tinder.
4. Once tinder ignites, blow gently to catch kindling on fire.
5. Build fire gradually by adding progressively larger sticks.

Tip: Magnesium shavings can be used if wet conditions prevent normal ignition.

MAINTENANCE

- Keep ferro rod dry
- Keep tinder dry
- Check striker edge for wear
- Replace cotton balls or waxed tinder after use
- Store kit in easily accessible location

TROUBLESHOOTING

No sparks

- Cause: Ferro rod worn, damp, or striker too smooth
- Fix: Dry rod, scrape clean, use proper metal edge

Tinder won't ignite

- Cause: Wet or compressed tinder
- Fix: Use dry cotton or waxed balls, fluff material

Fire extinguishes quickly

- Cause: Poor kindling, wind, or damp materials
- Fix: Prepare kindling in advance, shield fire, add dry fuel gradually

VARIATIONS

Pocket Survival Kit

- Small tin with cotton, ferro rod, striker, and magnesium
- Fits in pocket or survival bracelet

Backpack Emergency Kit

- Larger tin or waterproof bag

- Add mini candles, matches, and waterproof tinder

DIY Wax Tinder Cubes

- Mix dryer lint with wax in cupcake molds
- Quick ignition in wind or rain

BEST USE SCENARIOS

- Wilderness survival and bushcraft
- Camping trips and expeditions
- Off-grid emergencies
- Natural disaster survival kits
- Cold-weather survival
- Emergency signaling fires

PROJECT 11

DIY SIGNALING KIT (Emergency SOS & Rescue Signals)

SUMMARY

Being visible to rescuers can mean the difference between life and death. **A signaling kit** allows you to:

- Attract attention during wilderness survival
- Communicate distress in low-visibility conditions
- Signal for help from air or land rescue

This kit can be compact, portable, and effective in **day or night situations**, making it perfect for:

- Off-grid survival
- Hiking or mountaineering
- Flood or disaster zones
- Bushcraft and emergency kits

MATERIALS NEEDED

Core Materials

- Whistle (loud and durable)
- Mirror or reflective surface (for sunlight signaling)
- Bright-colored cloth or bandana (red, orange, or yellow)
- Flares or glow sticks (optional for night signaling)
- Waterproof container to store items

Optional Additions

- Signal flag (highly visible)
- Firestarter kit (for smoke signals)

- Battery-powered flashlight with SOS mode
- Extra rope or cord for raising signals

ASSEMBLING THE SIGNALING KIT

STEP 1 — Prepare Whistle
- Attach whistle to cord or keychain for quick access.
- Test whistle to ensure loud, clear sound (ideally 100+ meters).
- Keep dry and protected in waterproof container.

STEP 2 — Prepare Reflective Signaling
- Cut mirror or reflective card to 3–5 inches.
- Optional: Use polished metal or foil as substitute.
- Use for **sunlight signaling**: aim at rescuer or aircraft, flash repeatedly.

Technique:
- Hold mirror with one hand, cover a corner with finger, flash the reflected sunlight toward target.

STEP 3 — Prepare Visual Signals
- Fold bright-colored cloth or bandana.
- Can be tied to a branch or waved to attract attention.
- Optional: attach to tall stick or pole to make visible over distance.

STEP 4 — Night Signaling
- Glow sticks, battery-powered lights, or small flares can be included for nighttime SOS.
- Use light patterns like **3 flashes** (universal distress signal).
- Store lights in waterproof bag for emergencies.

STEP 5 — Assemble Kit
- Place whistle, mirror, bandana, flares, and other signaling items in **waterproof container**.
- Ensure compact size for portability in backpack, belt, or survival kit.

HOW TO USE THE SIGNALING KIT

Daytime

1. Use mirror to flash sunlight toward aircraft, rescuers, or boats.
2. Wave bright cloth or flag repeatedly.
3. Blow whistle in 3-blast pattern every few minutes.

Nighttime

1. Activate glow sticks or flashlight in SOS pattern.
2. Combine with whistle blasts for maximum attention.

Smoke Signals (Optional)

- Build small fire with damp leaves for white smoke.
- Create intermittent smoke bursts in 3 pulses for distress.

MAINTENANCE & SAFETY

- Keep whistle dry
- Polish mirror regularly for maximum reflectivity
- Replace glow sticks or flares after expiration
- Store kit in a known, accessible location
- Check bright cloth for fading or tears

TROUBLESHOOTING

Whistle not loud

- Cause: Moisture or debris inside
- Fix: Dry and clean, test before emergencies

Mirror flashes weak

- Cause: Dirty surface or poor angle
- Fix: Clean surface, adjust angle to sun

Cloth not visible

- Cause: Faded or camouflaged colors
- Fix: Replace with bright, high-contrast color

Night signaling fails

- Cause: Expired glow sticks or dead flashlight

- Fix: Test equipment regularly, replace as needed

VARIATIONS
Compact Pocket Version
- Small whistle, tiny reflective card, mini glow stick
- Fits in pocket or lanyard

Backpack Survival Kit Version
- Larger reflective panel, full-sized whistle, flares, rope, SOS flag
- For long-term bushcraft or off-grid expeditions

Urban Disaster Version
- LED flashlight with SOS mode, whistle, reflective tape, bright cloth
- Useful for floods, earthquakes, or blackout situations

BEST USE SCENARIOS
- Wilderness survival
- Off-grid living emergencies
- Flood, earthquake, hurricane aftermath
- Lost hiker or mountaineer situations
- Emergency kit for vehicles or boats

PROJECT 12

DIY TRAP & HUNTING KIT
(Basic Off Grid Food Procurement)

SUMMARY

When survival stretches past 48 hours, **food becomes important for strength, clarity, and long-term stamina**. A trap & hunting kit gives you **passive food acquisition** — meaning you can gather food while **conserving energy**.

This kit prepares you for catching:

- Small mammals (rabbits, squirrels, rats)
- Birds
- Fish
- Reptiles (depending on region)

Your goal here isn't to hunt like a pro — it's to **maximize results** with minimal movement, using basic materials and smart placement.

MATERIALS NEEDED

Core Items

- Paracord or strong cord (10–20 meters)
- Wire snare (20–30 gauge steel wire)
- Knife or multi-tool
- Small folding saw (for stakes and frame materials)
- Bait (nuts, seeds, berries, scraps)
- Survival whistle (for safety after trapping)
- Small pouch or waterproof bag

Optional Additions

- Fishing line + hooks
- Rubber bands or inner tube (for small slingshot)
- Small collapsible trapping net
- Carabiners
- Camouflage cloth or natural foliage

BUILDING THE TRAP & HUNTING KIT

SMALL GAME SNARE SETUP

This is your *bread and butter* for survival trapping.

Materials

- 20–30 gauge steel wire
- Branches or sticks
- Bait (if needed)

How to Build a Snare Loop

1. Twist one end of the wire into a small loop (about the size of a pen top).
2. Pass the other end of the wire through to form a slip-knot loop.
3. Size of loop depends on animal:
   - o **Rabbit:** 4–6 inches
   - o **Squirrel:** 3–4 inches
   - o **Rat:** 2–3 inches

Where to Place Snare

Target natural choke points:

- Small animal trails
- Low holes in bushes
- Water paths
- Tree bases with droppings

How It Works

Animal enters loop → wire tightens → no escape.

Pro Tip:

Camouflage the wire with dirt or mud.

SECTION 2

SPRING SNARE TRAP (For More Reliable Catch)

A more advanced snare with upward tension.

Materials

- Green sapling (acts as spring pole)
- Trigger mechanism (figure-4 or peg-style)
- Wire snare

How to Build

1. Bend flexible sapling.
2. Tie snare to end of sapling.
3. Position trigger so animal pulls bait → sapling snaps upward → hoists animal.

Best For

- Rabbits
- Squirrels
- Ground birds

SECTION 3

DEADFALL TRAP (Classic Survival Food Catcher)

This is the famous wooden trap where a heavy object falls on prey.

Materials

- Heavy stone/log
- Carved wooden triggers (figure-4 notch system)
- Bait

How to Build

1. Carve sticks into the classic "figure-four" shape.
2. Balance heavy object on trigger.

3. Place bait at the back.

4. When animal nudges bait, object falls instantly.

Best For

- Rats

- Mice

- Small scavengers

Warning:

Must be built carefully — test trigger several times.

SECTION 4

FISHING KIT (Low-Energy Food Source)

Fish are one of the highest return-on-energy survival foods.

Materials

- Fishing line (20–50 meters)

- Hooks (various sizes)

- Small weights

- Makeshift bobber (wood/bottle cap)

Methods Included in Kit

1. Still Line Fishing

Tie line to branch → hook + bait → leave overnight.

2. Trotline (For Bigger Yield)

Line stretched across water with multiple hooks attached.

3. Bottle Fish Trap

- Cut plastic bottle

- Insert cut end inverted

- Add bait

- Sink with stones

- Fish swim in but rarely escape

Best For: small fish, eels, river species.

SECTION 5

BIRD SNARE PERCH

A simple trap for catching ground-feeding or low-perching birds.

Materials
- Green branch
- Snare wire
- Perch stick

How It Works
1. Perch is set to collapse when bird lands.
2. Snare tightens around feet or neck.
3. Sapling pulls bird upward.

Effective For:
- Pigeons
- Small birds
- Ground feeders

SECTION 6

SLINGSHOT (OPTIONAL BUT EFFECTIVE)

A slingshot is low-weight, high-impact.

Materials
- Y-shaped branch
- Inner tube rubber or surgical tubing
- Leather patch for pouch

Targets
- Squirrels
- Birds

- Rabbits (short range only)

This is not for sport — strictly survival.

KIT ORGANIZATION

Pack all materials into a small waterproof bag, arranged like this:

Top Layer:
- Wire snares (coiled neatly)
- Fishing line + hooks

Middle Layer:
- Cord
- Smaller tools
- Pre-carved trigger pieces

Bottom Layer:
- Small saw
- Knife
- Pouch of bait

SAFETY GUIDELINES

Traps must **never** be placed near:
- Human trails
- Campsites
- Large predator paths
- Areas with high foot traffic

Always check traps every **6–12 hours** to reduce suffering and prevent trap predators from stealing your catch.

TROUBLESHOOTING

Snare not catching anything

- Wrong trail
- Loop too big/small
- Animals detecting human scent
→ Fix by observing tracks and wearing gloves.

Deadfall keeps collapsing early
- Trigger carved incorrectly
→ Smooth surfaces, correct notch angles.

Fishing line not biting
- Wrong depth or bait
→ Adjust location, test different baits.

BEST USE SCENARIOS
- Long-term wilderness survival
- Disaster aftermath when shops are closed
- Off-grid homestead food backup
- Forest, mountain, river, tropical, or bush regions
- Low-energy survival where hunting isn't viable

PROJECT 13

DIY WATER FILTRATION SYSTEM
(Advanced Off-Grid Water Purifier)

SUMMARY

Even when you have water sources, it's rarely safe to drink untreated water. A **DIY filtration system** removes:

- Dirt and sediment
- Microorganisms (bacteria, protozoa)
- Chemical impurities (partially, if using activated charcoal)

This system is ideal for:

- Long-term wilderness survival
- Flooded areas or post-disaster zones
- Off-grid living
- Emergency preparedness

Goal: Safe, drinkable water with minimal energy use.

MATERIALS NEEDED

Core Materials

- Large container or bucket (5–10 gallons)
- Smaller collection container or cup
- Sand (fine and coarse)
- Gravel or small stones
- Activated charcoal (from firewood or commercial)
- Cloth or coffee filter
- Knife or small scoop

- Optional: PVC pipe or funnel

Optional Enhancements

- Two-bucket setup for gravity filtration
- Ceramic filter cartridge
- Boiling pot for final sterilization
- Hand pump (for faster flow)

BUILDING THE WATER FILTRATION SYSTEM

STEP 1 — Prepare the Bucket/Container

- Use a clean, sturdy container.
- Drill or cut **small hole at the bottom** for filtered water to exit.
- Place **mesh or cloth over the hole** to prevent sand/gravel from falling out.

STEP 2 — Layer Materials

The layers filter water progressively:

1. **Bottom Layer (closest to outlet): Gravel / small stones**
   o 2–4 inches
   o Prevents sand from escaping
2. **Middle Layer: Sand (coarse then fine)**
   o 3–5 inches each layer
   o Coarse sand first, fine sand on top of coarse
   o Removes sediment, dirt, and debris
3. **Top Layer: Activated charcoal**
   o 2–3 inches
   o Adsorbs toxins, chemicals, and improves taste
4. **Optional Top Cloth Layer**
   o Prevents large debris entering filter
   o Acts as pre-filter

Tip: Ensure each layer is compact but not overly tight; water must flow freely.

STEP 3 — Assemble Two-Bucket System (Optional)

- Place large filtration bucket above collection bucket.

- Outlet hole directs filtered water into lower bucket.
- Gravity does the work; no pump required.

STEP 4 — Using the Filter

1. Pour raw water into the top layer slowly.
2. Collect filtered water from outlet.
3. Repeat filtration for cloudier water.
4. Boil filtered water if possible for **absolute safety**.

MAINTENANCE

- Replace charcoal every 1–2 weeks or after heavy use
- Rinse sand and gravel every few days
- Inspect bucket/container for cracks or leaks
- Keep top layer clean to prevent clogging
- Store filtered water in clean containers

TROUBLESHOOTING

Slow water flow

- Cause: Layers too compact or clogged
- Fix: Loosen sand, clean top layer, avoid overloading

Dirty water still coming out

- Cause: Large particles bypassing pre-filter
- Fix: Add cloth layer on top, clean sand and gravel

Water tastes odd

- Cause: Spent charcoal or contaminated materials
- Fix: Replace charcoal, clean system thoroughly

Leaks from container

- Cause: Hole too big or cracks
- Fix: Seal with waterproof tape or use replacement bucket

VARIATIONS

1. Portable Bottle Filter

- Small water bottle, sand, charcoal, small stones
- Gravity-fed or inverted bottle drip system
- Ideal for backpack survival

2. Ceramic or Commercial Cartridge

- Ceramic filter replaces charcoal/sand layers
- Can filter bacteria and protozoa efficiently

3. Multi-Stage Filtration

- Use solar still + DIY filter for ultimate purification
- First: Solar still for evaporation
- Second: Gravity filter with charcoal/sand

BEST USE SCENARIOS

- Off-grid wilderness survival
- Flooded or polluted water zones
- Emergency preparedness kits
- Camping in remote rivers or lakes
- Disaster aftermath in rural or urban areas

PROJECT 14

DIY SURVIVAL MEDICAL KIT
(First Aid Essentials Off-Grid)

SUMMARY

In survival situations, medical emergencies can happen anytime:

- Cuts, scrapes, or burns
- Fractures or sprains
- Insect bites or allergic reactions
- Illness from contaminated water or food

A **DIY survival medical kit** equips you to handle **minor to moderate injuries**, stabilize serious injuries, and prevent infection until help arrives.

This kit is portable, organized, and tailored for:

- Wilderness survival
- Off-grid living
- Disaster zones
- Bushcraft expeditions

MATERIALS NEEDED

Core Items

- Sterile gauze pads
- Adhesive bandages (various sizes)
- Medical tape
- Elastic bandages (for sprains)
- Antiseptic wipes or solution (alcohol, iodine, chlorhexidine)
- Tweezers and scissors

- Safety pins
- Disposable gloves
- Pain relievers (acetaminophen, ibuprofen)
- Antihistamines for allergic reactions
- Small emergency blanket

Optional / Advanced Items

- Splints (aluminum or wood)
- Burn gel or cream
- Antibiotic ointment
- Thermometer
- CPR mask or face shield
- Emergency oral rehydration salts (ORS)
- Digital pulse oximeter or small blood pressure cuff (for extended survival)

ORGANIZING THE KIT

Section 1 — Dressing & Wound Care

- Gauze pads
- Adhesive bandages
- Tape
- Antiseptic wipes
- Small scissors
- Tweezers

Section 2 — Medications

- Pain relievers
- Antihistamines
- ORS packets
- Burn cream or antibiotic ointment

Section 3 — Stabilization & Support

- Elastic bandages
- Splints
- Safety pins
- Emergency blanket

Section 4 — Protection & Hygiene

- Gloves
- Mask/face shield
- Hand sanitizer

Tip: Store all items in waterproof container or ziplock bag, clearly labeled, with compartments for quick access.

HOW TO USE THE SURVIVAL MEDICAL KIT

1. Wound Cleaning & Dressing

- Clean wound with antiseptic wipe
- Apply gauze pad or bandage
- Tape securely but not too tight

2. Sprains or Fractures

- Immobilize limb with splint
- Wrap with elastic bandage
- Apply cold pack if available to reduce swelling

3. Burns

- Rinse with cool water
- Apply burn gel or aloe
- Cover with sterile dressing

4. Insect Bites or Allergic Reactions

- Clean affected area
- Apply antihistamine cream or oral antihistamine
- Use elastic bandage if swelling is significant

5. Pain Management

- Administer acetaminophen or ibuprofen according to dosage
- Keep record of administration for repeated doses

MAINTENANCE

- Check expiry dates regularly
- Replenish used or worn items
- Keep kit dry and clean

- Rotate medications every 6–12 months
- Review kit contents periodically to ensure readiness

TROUBLESHOOTING

Gauze/medications wet

- Cause: Moisture or leaks
- Fix: Use waterproof container, double bagging

Bandages missing or torn

- Cause: Heavy use or poor storage
- Fix: Repack, carry spares

Pain relievers expired

- Cause: Over time
- Fix: Replace with fresh stock, mark expiry dates

Emergency kit disorganized

- Cause: Frequent use without repacking
- Fix: Use compartmentalized boxes or labeled bags

VARIATIONS

Compact Pocket Kit

- Small bandages, antiseptic wipes, pain relievers, gloves
- Ideal for hiking or short-term survival

Extended Survival Kit

- Full dressing, splints, burn care, medications
- For multi-day wilderness survival or disaster preparedness

Urban/Vehicle Kit

- First aid plus water purification tablets, flashlight, whistle
- Useful for floods, earthquakes, or stranded scenarios

BEST USE SCENARIOS

- Wilderness survival emergencies
- Off-grid living accidents
- Natural disaster zones (flood, earthquake, hurricane)
- Camping and bushcraft expeditions
- Vehicle or urban emergency kit

PROJECT 15

DIY SIGNAL FIRE & SMOKE SYSTEM
(Long-Distance Rescue Signals)

SUMMARY

When lost or stranded, being **seen or noticed by rescuers** is crucial. A **signal fire or smoke system** allows you to:

- Alert aircraft, boats, or distant teams
- Communicate your location over long distances
- Attract attention when other signaling methods fail

This system can be deployed in various environments, including:

- Forests
- Mountains
- Deserts
- Coastal or river areas

Goal: High-visibility signal that maximizes chances of rescue.

MATERIALS NEEDED

Core Materials

- Dry firewood or branches
- Tinder (cotton balls, dry leaves, or paper)
- Kindling (twigs, small sticks)
- Larger logs for sustained fire
- Rocks (to create fire ring or containment)
- Green leaves or moss (for smoke)
- Knife or multi-tool

- Lighter or ferro rod

Optional Enhancements

- Aluminum foil or reflective surface to increase visibility
- Cloth (bright colors) tied near fire for added contrast
- Metal grate or pot for controlled smoke production

BUILDING THE SIGNAL FIRE

STEP 1 — SELECT LOCATION

- Clear area of debris and dry grass to prevent uncontrolled wildfire
- Choose elevated or open area visible from distance
- Ensure wind direction carries smoke toward likely search areas

STEP 2 — CREATE FIRE BASE

- Build a fire ring with rocks for safety
- Ensure ground is stable and dry
- Keep water or sand nearby for emergency control

STEP 3 — BUILD FIRE STRUCTURE

- Small cone or teepee with tinder at center
- Surround with kindling, larger sticks, and logs
- Optional: Place green leaves or moss on top for dense smoke
- Arrange logs for slow, steady burn to prolong visibility

STEP 4 — IGNITE FIRE

- Light tinder with ferro rod or lighter
- Gradually add kindling to stabilize fire
- Add larger logs once fire is burning steadily

STEP 5 — CREATE SMOKE SIGNALS

- For maximum visibility, add green vegetation or moss
- Smoke bursts: 3 short puffs → universal distress signal
- Keep fire manageable and monitor constantly

Tip: Use wet leaves or moss intermittently for thick white smoke

- Dry wood creates mostly flames, less visible smoke

MAINTENANCE & SAFETY

- Never leave fire unattended
- Keep surrounding area clear of combustible material
- Use rocks or dug trenches to contain fire
- Keep water or sand nearby for quick extinguishing
- Monitor wind direction and adjust fire for visibility

TROUBLESHOOTING

Fire won't light

- Cause: Wet tinder, wind, or damp kindling
- Fix: Use dry tinder, shelter from wind, prepare backup tinder

Smoke too thin

- Cause: Dry wood only
- Fix: Add green leaves, moss, or grass

Fire spreads unintentionally

- Cause: Wind, dry surroundings
- Fix: Build fire ring, clear area, control logs, have water ready

Visibility low

- Cause: Fire in depression or behind obstacles
- Fix: Move fire to higher ground, add reflective cloth or signal mirror

VARIATIONS

Portable Cone Smoke Signal

- Small cone of dry wood + green moss in metal container
- Can be carried and deployed quickly

Daytime vs Nighttime Signals

- Day: Focus on smoke density
- Night: Focus on flames with bright reflective materials for contrast

Urban/Disaster Use

- Use fire or smoke in safe outdoor open areas
- Combine with bright-colored cloth or reflective panels

BEST USE SCENARIOS

- Wilderness survival
- Mountain or desert emergencies
- Flood or hurricane aftermath
- Lost or stranded hikers
- Remote river or coastal locations

PROJECT 16

DIY OFF-GRID COOKING SYSTEM
(Bushcraft & Survival Meals)

SUMMARY

In survival scenarios, cooking food safely is crucial to:

- Prevent illness from bacteria or parasites
- Maximize nutrition from foraged or hunted food
- Boil water for drinking or hygiene
- Maintain energy and morale

An **off-grid cooking system** uses natural materials, low-tech methods, and minimal fuel to prepare meals in the wilderness or disaster zones.

This system is ideal for:
- Bushcraft and camping
- Wilderness survival
- Disaster preparedness
- Off-grid homesteads

MATERIALS NEEDED
Core Materials
- Knife or multi-tool
- Small folding saw
- Firestarter kit (from Project 10)
- Large, heat-safe container (metal pot, tin can, or clay pot)
- Tripod or makeshift stand for pot

- Rocks (for heat insulation and stability)
- Water source
- Fuel (wood, dry leaves, twigs, charcoal)

Optional / Advanced Items

- Grate or wire mesh for grilling
- Clay or stone oven for baking
- Solar cooker (reflective panel + black pot)
- Rope or paracord (to suspend cooking pots)

BUILDING THE OFF-GRID COOKING SYSTEM

SECTION 1 — SIMPLE CAMPFIRE COOKING

1. Prepare fire pit (see Project 16 for fire safety).
2. Build small teepee or log cabin fire structure.
3. Place pot or container over fire using:
   - Tripod made from sticks and rope
   - Flat stones as stove platform
   - Forked stick support
4. Boil water or cook food directly over flames.

Tip: Use small logs or twigs for consistent, controllable heat.

SECTION 2 — STONE HEAT BED COOKING

1. Heat large flat stones in fire for 20–30 minutes.
2. Remove fire and spread stones as cooking platform.
3. Place food directly on hot stones or wrap in leaves for roasting.

Best for: Meat, root vegetables, fish

SECTION 3 — CLAY OR PIT OVEN (Baking/Slow Cooking)

1. Dig shallow pit in safe area.
2. Line pit with stones, then add hot embers.
3. Place food in clay or wrapped in leaves on embers.
4. Cover with more embers and soil to trap heat.
5. Cook 1–3 hours depending on food type.

Best for: Root vegetables, meat, or whole fish

SECTION 4 — SOLAR COOKER (Optional)

1. Use reflective panel (aluminum foil, mirror, or reflective tarp)
2. Direct sunlight to black pot or container with food
3. Secure pot for stable heating
4. Cooking times: 1–4 hours depending on sun intensity

Advantages: Fuel-free, smoke-free, safe in dry areas

SECTION 5 — IMPROVISED GRILL OR SKEWER COOKING

- Use long green sticks as skewers for meat or fish
- Use wire mesh or grate over fire for grilling
- Rotate food for even cooking
- Keep flames moderate to avoid burning

MAINTENANCE & SAFETY

- Always monitor fire while cooking
- Keep water nearby for emergencies
- Use heat-resistant gloves or cloth when handling hot pots
- Keep food off the ground to avoid contamination
- Ensure proper ventilation if cooking in enclosed shelters

TROUBLESHOOTING

Food not cooking evenly

- Cause: Uneven heat or fire too low
- Fix: Adjust firewood, rotate pot or food

Pot tipping or unstable

- Cause: Improper tripod or stones
- Fix: Reposition support, ensure stability before placing food

Fire goes out quickly

- Cause: Wet wood or insufficient tinder
- Fix: Use dry kindling, prepare backup fuel

Smoke too heavy

- Cause: Wet wood or excessive fire materials
- Fix: Use dry wood, ventilate cooking area

VARIATIONS

Portable Cooking Kit

- Small pot, folding stove, or tin can stove
- Ideal for backpack survival

Pit Oven System

- Slow-cooking meals for multi-day survival
- Requires minimal attention after setup

Solar Cooker

- Fuel-free option in sunny climates
- Ideal for emergency cooking without fire

BEST USE SCENARIOS

- Wilderness survival expeditions
- Off-grid living or homesteads
- Disaster aftermath (flood, hurricane, earthquake)
- Camping and bushcraft adventures
- Emergency food preparation without electricity or gas

SECTION 5

BONUS MINI PROJECTS

MINI PROJECT ONE
EMERGENCY FIRE STARTER POUCH

Summary: Quick-access pouch for fire-starting in emergencies.

Materials:

- Cotton balls or dryer lint
- Petroleum jelly or beeswax
- Small waterproof container or ziplock bag

Instructions:

1. Coat cotton balls or lint in petroleum jelly or melted beeswax.
2. Store in waterproof container.
3. To use, pull one out, light with ferro rod or lighter. Burns 5–10 minutes reliably.

Best Use: Quick fire start when natural tinder is wet or scarce.

MINI PROJECT TWO
PORTABLE WATER PURIFICATION
DROPS

Summary: Quick chemical purification method.

Materials:

- Water purification tablets or chlorine drops
- Small container or sachets

Instructions:

1. Carry small tablet or drop bottle in kit.
2. Add to collected water per dosage instructions.

3. Wait required time (usually 30 minutes).
4. Water is safe to drink.

Best Use: When boiling or filtering isn't immediately possible.

MINI PROJECT THREE
DIY PARACORD BRACELET WITH SURVIVAL TOOLS

Summary: Wearable survival kit in a bracelet.

Materials:

- 10–15 ft paracord
- Small whistle, firestarter, tiny blade or fishing hooks
- Buckle or knot closure

Instructions:

1. Weave paracord into bracelet, integrating small tools.
2. Can unravel to use full cord in emergencies.

Best Use: Hands-free kit, emergency rope, signaling, or first aid.

MINI PROJECT FOUR
IMPROVISED CAMOUFLAGE SHEETS

Summary: Quick hide or shelter enhancement.

Materials:

- Lightweight tarp, cloth, or poncho
- Mud, leaves, or foliage

Instructions:

1. Cover tarp or cloth with mud, leaves, or other natural materials.
2. Use to hide gear, yourself, or your campsite.

Best Use: Emergency concealment, wildlife observation, or stealth survival.

MINI PROJECT FIVE
MAKESHIFT TOOLS & UTENSILS

Summary: Create usable tools from natural materials.

Materials:

- Branches, vines, or stones
- Knife or multi-tool

Examples:

1. Wooden spoon or spatula for cooking
2. Sharpened stick for digging or hunting
3. Rock hammer for cracking nuts or shaping wood

Best Use: Sustained survival when commercial tools are unavailable.

MINI PROJECT SIX
SIGNALING WHISTLE KNOT

Summary: Enhance whistle visibility and usability.

Materials:

- Small whistle
- Cord, string, or paracord

Instructions:

1. Attach whistle to cord.
2. Tie in loop or around neck for quick access.
3. Can use in combination with mirrors or fire for SOS signaling.

Best Use: Quick, hands-free emergency alerts.

MINI PROJECT SEVEN
DIY EMERGENCY FOOD WRAPS

Summary: Preserve and cook small portions of food.

Materials:

- Large leaves (banana, palm, or plantain)
- Stones or sticks for weighing down

Instructions:

1. Wrap food in large leaves.
2. Can place near fire or hot embers for cooking.
3. Keeps food clean, portable, and prevents contamination.

Best Use: Short-term storage or cooking when cookware is limited.

MINI PROJECT EIGHT
PORTABLE SUN DRYING RACK

Summary: Preserve food for later using sunlight.

Materials:

- Lightweight sticks or small frame
- String, mesh, or cloth

Instructions:

1. Stretch mesh or cloth on frame.

2. Place thinly sliced meat, fruits, or vegetables.
3. Leave in sun to dry. Cover with mesh to avoid insects.

Best Use: Creating emergency food rations in off-grid survival.

MINI PROJECT NINE
EMERGENCY SIGNAL MIRROR COMPACT

Summary: Carryable reflective signaling tool.

Materials:

- Small mirror, polished metal, or CD fragment
- Lanyard or small case

Instructions:

1. Attach to lanyard or kit.
2. Use sunlight to flash signals for aircraft or rescuers.

Best Use: Lightweight signaling in daytime when other methods aren't practical.

MINI PROJECT TEN
QUICK IMPROVISED SEAT OR PAD

Summary: Protect from cold or wet ground during rest.

Materials:

- Large leaves, branches, or moss
- Waterproof tarp (optional)

Instructions:

1. Lay natural materials or tarp on ground.
2. Sit or kneel for comfort and insulation.

Best Use: Rest, injury treatment, or food preparation on cold/wet surfaces.

Maintenance & Care Guide for Survival Gear

Objective: Ensure all tools, kits, and materials last longer and stay functional.

Fire & Lighting Equipment

- Keep fire-starting materials dry. Store in waterproof containers.
- Test ferro rods and lighters monthly.
- Replace fuel in oil lamps regularly.

Water Filtration & Purification

- Clean sand, gravel, and charcoal layers periodically.
- Replace charcoal every 1–2 weeks or after heavy use.
- Boil filtered water when possible to ensure complete safety.

Traps & Hunting Gear

- Check snares and traps every 6–12 hours to prevent spoilage and unintended catches.
- Replace worn or rusted wires promptly.
- Camouflage traps with natural materials to maintain efficiency.

Medical Kit

- Replace expired medications.
- Keep dressings dry and sterile.
- Reorganize kit after every use.

Cooking & Shelter Materials

- Store tarps, ropes, and containers in dry areas.
- Inspect and repair shelter structures periodically.
- Keep cooking utensils and pots clean to prevent contamination.

Survival Tips & Best Practices

- **Prioritize safety first:** Avoid unnecessary risks; always have an exit plan.
- **Energy management:** Conserve energy by using passive food acquisition, traps, and gravity-fed water systems.
- **Observe nature:** Animal trails, wind direction, and sunlight patterns help in locating resources and navigation.
- **Signal wisely:** Use mirrors, whistles, and smoke signals in bursts for maximum visibility.
- **Practice skills:** Train with your survival kits before relying on them in an actual emergency.

Emergency Checklist (Quick Reference)

Always Carry:

- Water purification method (filter or tablets)
- Fire-starting kit
- Knife/multi-tool
- First aid kit
- Signaling tools (whistle, mirror, flashlight)
- Compact food rations or bait for traps

Optional Items:

- Paracord bracelet
- Solar-powered light
- Improvised cookware
- Survival blanket or tarp

Notes & Record-Keeping

- Keep a small **survival journal**:
  - o Record trap locations and efficiency
  - o Note water sources and filtration success
  - o Track food supplies and cooking results
  - o Document lessons learned for continuous improvement
- Helps in **planning long-term survival strategies** and adapting to changing conditions.

Author's Final Notes

For over a decade, I have served in a tactical military unit specializing in field operations, emergency logistics, and outdoor survival. With countless hours spent in challenging environments, from dense forests to conflict zones, I learnt the practical realities of surviving without comfort, electricity, or external support. After leaving active duty, I shifted my focus to training civilians through hands-on workshops and disaster-preparedness courses. My mission is simple: teach everyday people the kind of skills that save lives when everything else fails.

* 9 7 8 1 9 5 6 3 6 9 3 0 4 *